Building a

"REAL"

Vertical Steam Engine

by

Andrew Smith

Model and Allied Publications
Argus Books Ltd.
14 St. James Rd., Watford
Herts, England

in association with

Henley Engineering Publications Ltd.
54 Berkshire Rd.,
Henley-on-Thames
Oxon

© *Andrew Smith and Henley Engineering Publications Ltd.*
1977

ISBN 0 85242 524 4   Argus Books Ltd.
ISBN 0 905180 04 6   Henley Engineering Publications Ltd.

First published 1977

Printed in England by Higgs & Co., Caxton House, Henley-on-Thames

Building

a

"REAL"

Vertical Steam Engine

BY

Andrew Smith

N.B.—Do not scale from drawings

The completed "REAL" Vertical Steam Engine

Preface

Although the Stuart Turner series of model steam engines covers a very wide range of types, there is always the interesting possibility of developing a new configuration by making use of existing castings. The accuracy of dimension and easy working for which these castings are noted makes such a choice even more attractive in a hobby where "one off's" tend to be difficult to obtain. Such "variations" serve a number of useful purposes; they form an interesting armchair and drawing board exercise and by increasing the use of such castings make their manufacture more economic at a time when any new production is expensive.

The proportions of the cylinder, flywheel and other castings used in this series of engines, belongs to a period when there seemed to be a limitless range of variations in steam engine styles. Apart from the Beam and the Horizontal, there was the Table Engine, the Grasshopper, the Side Lever, the Bell Crank and the Vertical. In fact, the type of engine that today we refer to as the Vertical would, in those days, have been considered an Upside-down Vertical!

I hope that you will enjoy building and steaming these engines; they will not power your workshop, but I am certain will reward you with a great deal of pleasure and quiet contentment.

ANDREW SMITH.

Saltford,
    Avon,
        England.

1. Introduction

In the preface, comment was made about "upside down" vertical engines. Perhaps a more erudite description would have been to refer to them as "inverted vertical engines." The term does, in fact, apply to the form of engine common among those models in the Stuart Turner catalogue, with the crankshaft below and the cylinder mounted above it. The original and "true" vertical engines having these components in the reversed positions.

The beam engine of the late 18th and early 19th centuries was ideally suited to work requiring a reciprocating form of motion as in pumping water from mines, which was their initial and prime purpose. When the desire for a rotative motion appeared, for the driving of machinery, the existence of a massive and heavy beam between the reciprocating action of the piston and the rotation of the crankshaft was soon seen to be an unnecessary evil, especially in the smaller engines. No doubt some enterprising mechanic soon realised that if he removed the heavy beam and superimposed the crankshaft above the cylinder, he could couple the piston rod to the crank by means of a connecting rod and the mechanism would work efficiently. The fact that he sold the heavy cast-iron beam to the local scrap dealer was probably overlooked by the owner of the manufactory when the aforesaid mechanic demonstrated how much faster the engine would work when freed from the sluggish inertia of the heavy beam.

Before long, such vertical engines were being built in their own right. In the smaller sizes especially, their simplicity and the small space they occupied made them popular in the smaller factories; although very large examples, with cylinders of over five feet bore, were made as colliery winding engines. However, the example we have chosen is based on one of 8 inch cylinder bore described in "The Stationary Steam Engine" by George Watkins. Its colliery work was equally as important as the enormous pumping and winding engines because it drove the plant which prepared the fodder for the pit ponies! It is likely to have been made about 1850 and was dismantled almost exactly a hundred years later.

In honour of the unknown "rude mechanic" who had the wit to realise that the beam was an unnecessary anachronism and conceived the first vertical engine, I have called this a "Real" vertical engine, but perhaps, because of the duties it carried out faithfully for a whole century, we should call it the "Pony" Engine; I leave the choice to you.

The work entailed in this engine is simple and straightforward. In fact it would form, for the newcomer, an ideal follow-on from the simple "VICTORIA" and form an interesting comparison as they are both engines of a type used for driving small workshops during the latter half of the last century. Such power units were often produced by local blacksmiths who were far-seeing enough to realise that the horse, which had been their maintstay for generations, would give way to mechanical sources of power.

Turning the columns

2.   Engine base and structure

To start work on building this period type engine, I would suggest that we commence with the columns. As there are four of these and they are a fair length and to be machined all over, they will take a considerable amount of time to finish. Don't let this put you off, because the turning is simple between "centres work" and with them out of the way, the biggest single bit of machining will be complete.

They are turned from $\frac{5}{8}$ inch diameter bright drawn mild steel bar and I have asked Stuart Turner Ltd. if they will be so kind as to supply this to the correct overall length and with the ends faced and centre drilled. This will be a distinct advantage to builders whose lathes are limited in the length they will accept between the headstock and tailstock centres.

If, however, you are building from scratch, you will have to face and centre the bar ends yourself. If your lathe will take $\frac{5}{8}$ inch diameter through the headstock spindle, e.g. the ML7, this initial bit of work will be easy. If you have a less accommodating lathe (like mine), use a fixed steady to support the end of the bar on which you are operating while the other end is held in the self-centring chuck. If you have no fixed steady, rig up a $\frac{5}{8}$ inch bore bush of some sort, clamped, at centre height, to the lathe bed. Even a block of hardwood with a $\frac{5}{8}$ inch hole bored through will do at a pinch.

As a last resort, the job can be done by filing the ends of the bar square, then find the centre either with a centre square or a pair of odd-leg calipers, centre punch and drill in the drilling machine or by hand. If you are doing it this way, use a suitable size twist drill to produce the initial hole and finish with a centre drill—the mortality rate among centre drills is extremely high and they are very expensive.

Having prepared the ends of each of the four lengths of bar, mount them between centres in the lathe and reduce the columns to $\frac{1}{2}$ inch diameter leaving a shoulder of full diameter at each end. A front turning tool with the nose rounded to 1/16 inch radius will not only efficiently remove the metal but also leave the desired radius at each end. The final cut should be only a very light one, with a fine feed and a plentiful supply of cutting oil. It is no waste of time to spend a few minutes dressing the tool edge with an oilstone slip, just prior to taking the final cut. Do not, during this particular turning operation, reduce the ends of the bars, but leave them for the moment, the full $\frac{5}{8}$ inch diameter.

We deal with this next operation now. Swop the round nosed lathe tool for a knife tool and reduce the bar ends to 0·185 inch diameter for a length of 15/16 inch at one end and 5/16 inch at the other. The length of the columns between shoulders is to finish at $9\frac{3}{4}$ inches. Thread the reduced ends 2BA keeping the die as square as possible so that you do not get a drunken thread. The position of the 5/16 inch wide by 1/16 inch deep recess which supports the piston rod guide stay, may be lightly marked out at this stage, but do not file it out until the stay has been made.

Before we continue further, a few words about drill sizes in view of the transition to metric standards.

TAPPING AND CLEARING SIZE DRILLS

May I draw your attention to this list of tapping and clearing drill sizes. The beginner may be surprised to see that, for example, he could choose any one of ten different drills as tapping size for a 5BA screw. In fact, the difference in diameter between the first and last of these listed is only $8\frac{1}{2}$ thousandths of an inch, that is just twice the thickness of the paper on which this book is printed. Why the list? Well, due to metrication, number and letter size drills are being phased out and eventually Imperial sizes as well, but most of us will be using those we have for, I hope, a long time yet. If you are a beginner, choose a drill at the bottom of the list and you will have no difficulty in achieving successful tapping. If more experienced, work about the middle of the range and you will get a thread of fuller form. If you use a drill at the top of the list, then for goodness sake keep a steady hand; I have just paid over £1 for a high speed steel 7BA tap!

Screw size	Tapping size drill	Clearing size drill
2BA	No. 26	3/16 inch
	No. 25	4·80 mm
	3·80 mm	No. 12
	No. 24	No. 11
	3·90 mm	4·90 mm
	No. 23	No. 10
	5/32 inch	
	No. 22	
	4·00 mm	

4BA	No. 34	No. 27
	2·85 mm	3·70 mm
	No. 33	No. 26
	2·90 mm	
	No. 32	
	2·95 mm	
	3·00 mm	
5BA	No. 40	No. 30
	2·50 mm	3·30 mm
	No. 39	3·40 mm
	2·55 mm	No. 29
	No. 38	
	2·60 mm	
	No. 37	
	2·65 mm	
	2·70 mm	
	No. 36	
6BA	No. 44	No. 34
	2·20 mm	2·85 mm
	2·25 mm	No. 33
	No. 43	2·90 mm
	2·30 mm	No. 32
	2·35 mm	
	No. 42	
7BA	No. 48	No. 39
	1·95 mm	2·55 mm
	5/64 inch	No. 38
	No. 47	2·60 mm
	2·00 mm	No. 37
	2·05 mm	2·65 mm
	No. 46	2·70 mm
	No. 45	No. 36

With the columns complete, the base and entablature can be worked together, as the machining required is similar on each piece. These are new castings which I hope Stuart Turners may see fit to help us out with.

The underside of the base may be drawfiled flat and true, then gripped in the four-jaw chuck and, at slow speed, a light cut taken over the top surface. The entablature can be faced, top and bottom, in the lathe to finish ¾ inch thick. If the castings are to S.T.'s usual quality, they will both probably only require cleaning up with a smooth file.

The position of the 2BA clearing and 2BA tapped holes may now be marked out, centre punched and completed. The clearing holes, in each casting, take the four columns and must therefore line up with each other, so drill them first with a pilot drill and offer them up to each other to see that they do so. It might be sensible to clamp the two parts together when finally opening out with the 2BA clearance drill to ensure accuracy. Make your choice of drill sizes from the table earlier in the book.

9

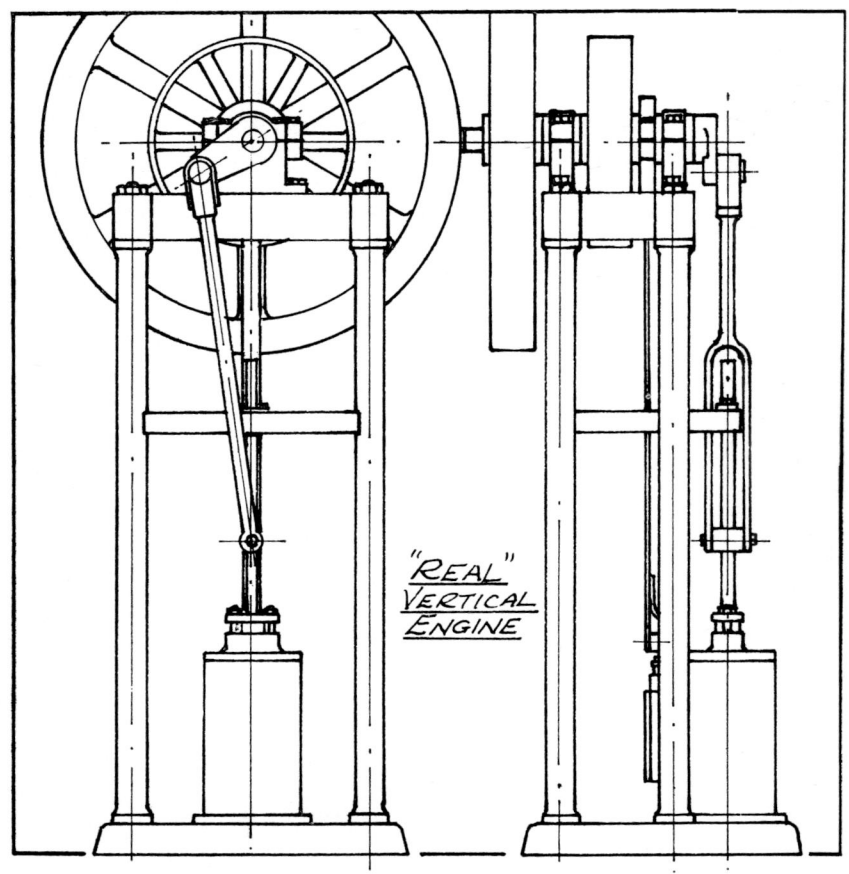

"REAL"
VERTICAL
ENGINE

Because I have imagined this engine as a free standing model, some liberties have been taken with the design of the entablature by making it a self-contained unit. In reality, and as shown in Mr. Watkin's book, this might consist of cast-iron beam sections of such a length that their ends were built into the engine house wall. If you prefer this approach, and intend a scenic type of model, you may machine or build-up replicas of these cast-iron girders from mild steel, instead of using the entablature casting.

While on this theme, you will probably notice different patterns of entablature shown in the photograph of my prototype engine and some of the drawings. This is because at the present time of writing, we are still undecided as to which style (a) is most easily machined; (b) looks best; (c) moulds conveniently. The version on my original engine was made from a scrap of $\frac{3}{4}$ inch thick steel plate by drilling, hacksawing and filing!

Assembly of the engine structure is easy. If any lack of squareness exists, it will be due *either* to the 2BA clearance holes not lining up with

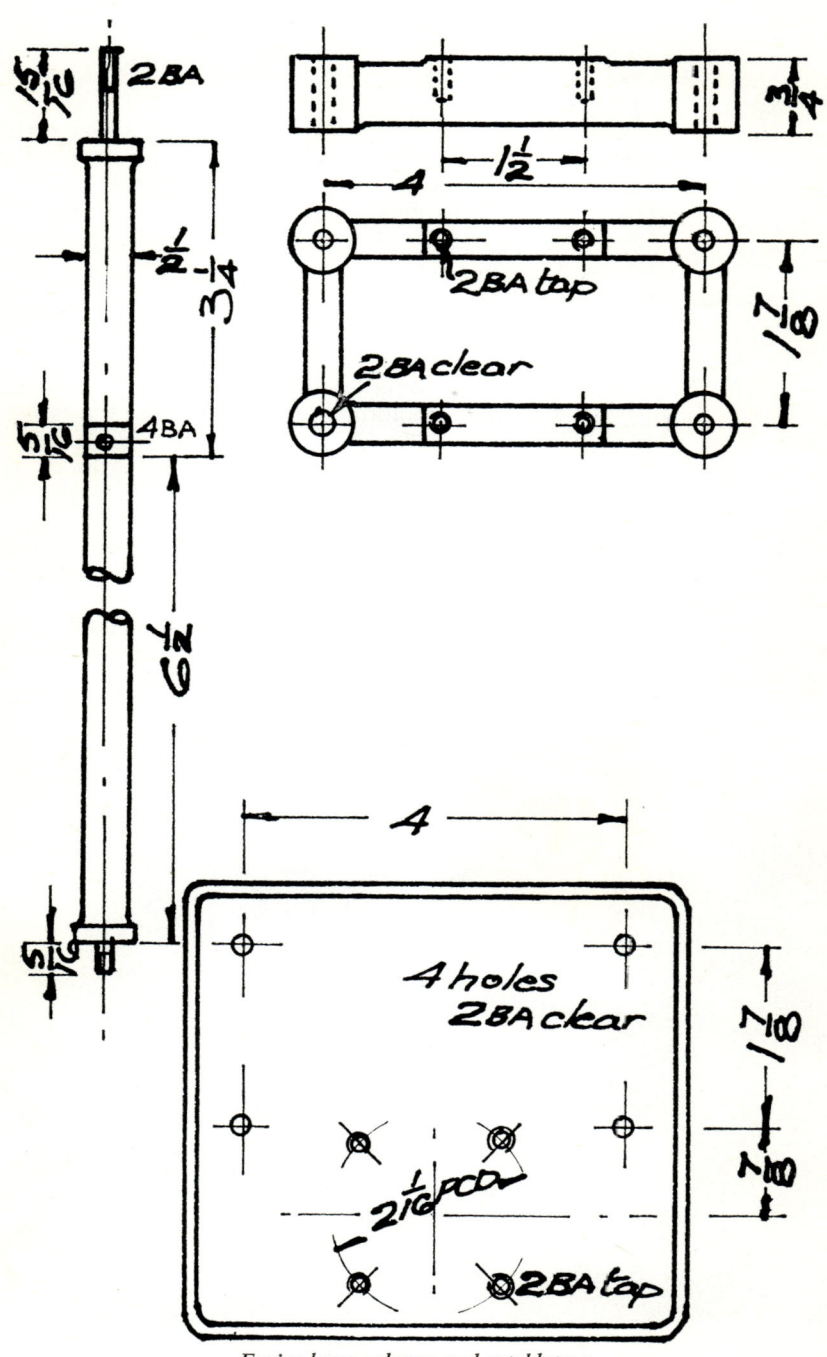

Engine base, columns and entablature

11

each other—in which case open up the offending hole until the matter is put right—*or* to the columns not being exactly the same length between shoulders, when the remedy is to carefully face off the excess until they all match.

It will no doubt be the builder's intention to mount the engine on a wooden or other base. On the drawings no lugs or holes are shown for this. I would suggest three 2BA clearance holes for this pupose, one at each of the front corners of the base and one in the centre of the rear edge.

3. Crankshaft, bearings, flywheel and pulley

With the engine structure complete, it would be nice to get some of the moving parts made; but first, the main bearings need to be fitted. These are in two parts, top and bottom castings in gunmetal, a copper, tin, zinc, alloy which machines beautifully and has excellent wearing qualities.

Clean the castings all over with a file, then mount in the four-jaw chuck and machine the mating faces on each of the four parts. Paint these faces with solder paint and heat until it melts over the surfaces, or tin using stick solder and flux. Clamp the pieces together and reheat until the pairs are sweated together. Allow to cool and wash to remove excess flux which is normally corrosive.

Mount each bearing in the four-jaw chuck, base outwards, and face off until the distance from joint to underside is $\frac{7}{8}$ inch. Remount in an upright position in the chuck with the bearing centre running true.

Main bearing set up in the four-jaw chuck, about to be reamed

Face, then centre drill, open out and ream to 7/16 inch or finish to size with a small boring tool if a reamer is not available. With a knife tool face down the side of the bearing block leaving a protruding rim 11/16 inch diameter and $\frac{1}{8}$ inch long. Slacken two chuck jaws only—not the one supporting the bottom of the bearing—and reverse the bearing. Reset, so that the 7/16 inch bore is running reasonably true, face to an overall thickness of $\frac{5}{8}$ inch, then reduce the side of the bearing to $\frac{3}{8}$ inch thick leaving a rim 11/16 inch diameter by $\frac{1}{8}$ inch long. Remove from the lathe and clean up the remainder of the bearing to size by careful filing. Mark out for the 2BA mounting holes at $1\frac{3}{8}$ inch centres, pilot drill and finish with a 2BA clearance drill. On the top cap of the bearing, mark out for the 5BA bolt holes, drill tapping size to a depth of $\frac{3}{8}$ inch, and spot-face the top surfaces. Now drill No. 40 and tap $\frac{1}{8}$ inch by 40 tpi for the fitting of oil cups (part No. 183/1), in the centre of the top bearing cap. Finally, identify the pairs of bearings and unsweat them; clean off the solder, tap the lower halves of the clamping bolt holes 5BA and open out the bolt holes in the cap with a 5BA clearance drill.

Spot through from the bearings for the 2BA holes on the top of the entablature. Drill them tapping size and tap 2BA. Put the crankshaft or a piece of 7/16 inch diameter rod through the bearings as you bolt them down to ensure that they line up correctly with each other. If they are a bit stiff when fully tightened down, pass a 7/16 inch reamer through them, but only to correct slight inaccuracies.

The crankshaft, which comes next, is in three parts, the shaft, the crankweb and the crankpin. As the first and the last have to fit accurately into the crankweb I suggest that we make that item first. This is a neat little malleable iron casting which initially requires cleaning up with a file to its required shape and size. Now set it in the four-jaw chuck with the flat side outwards and face off making sure that you leave sufficient to allow for finishing to the final thickness. Reverse in the chuck and set the $\frac{3}{4}$ inch diameter boss to run true. Face to $\frac{3}{8}$ inch thick, centre drill and drill and ream to $\frac{3}{8}$ inch. While the casting is still mounted in the chuck, advance the tool and face down the small boss to $\frac{1}{4}$ inch thick. Remove from the lathe, mark out, centre punch, drill and ream the 3/16 inch hole in the smaller boss at 1 inch centres. If you do not have a drilling machine which is sufficiently accurate, I suggest that you do this job with the crankweb clamped to the lathe faceplate and running true. Prior to a reaming operation use a drill as near to the reaming size—but obviously below it—as possible. For example, for the 3/16 inch reamed hole, the drill would ideally have been No. 13 or 4·70 mm—which are actually both the same size—leaving only 2·5 thousandths of an inch to be removed by the reamer. If you should not have the correct size of reamer, carry out the same procedure replacing the reamer with a drill, preferably a new one.

The two parts which have to fit the crankweb are simple turning jobs, requiring the self-centring chuck and a keen knife tool. Mount the $\frac{3}{8}$ inch diameter bar in the chuck with sufficient protruding to make the crankpin in one setting. In this way we can be sure that all diameters are turned concentric with each other. Face the end, then turn down a $\frac{7}{8}$

13

Drilling the crankweb prior to reaming

inch length to just over $\frac{1}{4}$ inch diameter. Reduce $\frac{1}{2}$ inch length to 3/16 inch diameter aiming for a very close fit in the crankweb. Now reduce 5/16 inch of this length to 0·185 inch diameter and thread 2BA, guiding the diestock with the tailstock barrel. When the thread is complete, shorten the length of this part to the 7/16 inch shown on the drawing, leaving a nicely radiused finish to the end of the thread. Reduce the remainder to an accurate $\frac{1}{4}$ inch diameter with as good a surface finish as possible. Finally, part off, leaving a head 1/16 inch thick.

If the material for the crankshaft is supplied as oversize stock it should be turned between centres. Start by holding the bar in the three-jaw chuck for facing the ends and centre drilling with a small centre drill preferably a No. 1 size. Then replace the three-jaw chuck with a catch-plate and clamp a small lathe carrier on the work. Using a slow feed turn the bar to 7/16 inch diameter; use the main bearings as a gauge. Reverse the bar in the centres and turn the unmachined end down to approximately $\frac{3}{8}$ inch diameter, a light press fit in the crankweb. Go carefully here, because too tight a fit may result in a broken crankweb or a bent shaft, if too much force is applied when pressing them together, while a sloppy fit cannot be rectified and will mean starting again and shortening the shaft.

When the shaft has been pressed home, drill a cross hole No. 42 and fit a 3/32 inch pin. File each end of the pin down flush with the web so that it is not visible.

The first operation on the flywheel is to clean all over with a smooth file to remove parting line flashes from the casting process. There are different ways by which the machining of the flywheel may be accomplished on a small lathe, but I suggest the most convenient method is to use the faceplate for all the operations. Clamp the casting to the faceplate with bolts between the spokes and suitable thicknesses of packing between the faceplate and the spokes so that the latter run no risk of distortion. Set the wheel to run as truly as possible; then with the lathe set on its slowest speed, face the side of the rim and its periphery. Speed up the lathe and face the wheel boss. Centre drill the boss, open out and ream or preferably bore to 7/16 inch. I say bore preferably because this hole and the outer rim must be concentric—nothing looks worse than a flywheel that wobbles! Now reverse the casting on the faceplate and face the opposite side of the rim to $\frac{3}{4}$ inch wide and likewise the boss to 1 inch long. With a smooth flat file—complete with handle!—remove the sharp edges and the flywheel is complete apart from drilling and tapping for the 2BA grub screw.

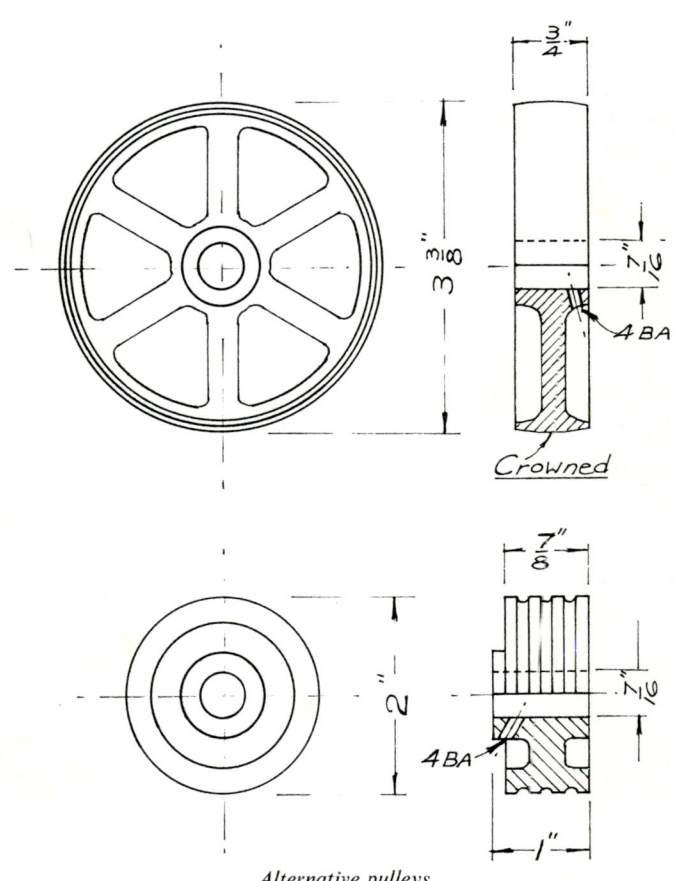

Alternative pulleys

The turning of the driving pulley is a similar series of machining operations to those just dealt with. It may be accomplished on the face-plate or done by gripping the boss of the pulley in the self-centring chuck. Face the side of the rim; then turn the periphery. As a pulley for a flat belt has to be 'crowned'—that is the centre must be higher than the sides to stop the belt from running off, we cannot get away with leaving the periphery of the pulley flat. To accomplish this set the topslide over by a very small amount—no more than a couple of degrees—and with the handle end of the slide away from you. Now take a cut across the out-side of the pulley from the right hand edge and running out at the centre. All the rest of the operations are as for the flywheel but note that to finish the other half of the crowning operation, the pulley will have to be mounted on a mandrel—we can't trust the accuracy of a so-called self-centring chuck. Blend the two sides of the pulley crown into each other with a dead smooth file.

Assemble the crankshaft and wheels on to the engine structure, spin them round, and watch, with delight, how that large flywheel rotates with not a waver!

4. The steam cylinder

The cylinder assembly for this engine of the series is almost a replica of that for the "STUART" beam engine, the difference being in the length of the piston rod which in this vertical engine is fractionally over 7 inches long. This is necessary, because the piston rod itself is used as a form of crosshead guide; a simple technique used by many small builders to obviate the need for long planed surfaces.

I keep commenting on the quality of the cylinder casting used in this range of historical engines, but it is a fact that the production of cast-in steam ways in such a small cylinder is a work of art and a great boon to the model engine builder whose skill or equipment is limited. In passing, I have just been looking at a photograph of the "Rocket" in the Science Museum in London and noticed how similar its cylinders are to the type we are using; after all it is about the same period in steam engine development. The cylinders were of 8 inch bore and 17 inch stroke so a pair of the Stuart Turner cylinders could form the basis of a delightful $7\frac{1}{4}$ inch gauge model—daydreaming again!

I held the cylinder casting in the four-jaw chuck with the port-face outwards. The chuck is the $4\frac{1}{2}$ inch light type (Burnerd) which, I imagine, is about the smallest anyone will be using for this work. The cylinder located in the chuck jaw channel which acted like a vee block to help steady it, and the distance from the chuck face to the cylinder centre-line was assessed. This dimension, added to $\frac{7}{8}$ inch gave the rule measure-ment at which facing of the port face had to stop. Just over 1/32 inch had to be removed: two very light cuts and the job was done. As this method of mounting the cylinder had been so successful, two jaws were slackened and the cylinder rotated 90° about its longitudinal axis, bringing the exhaust flange face outwards, and this machined.

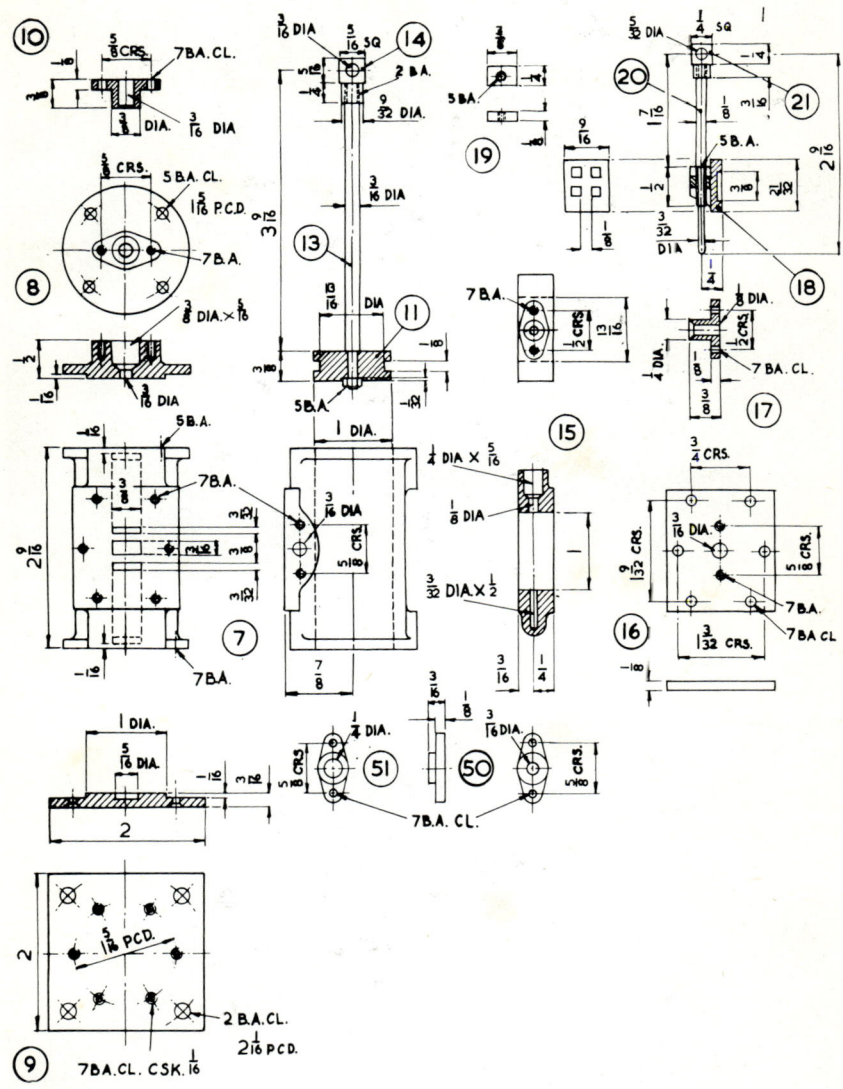

Cylinder and associated components

17

The portface just completed

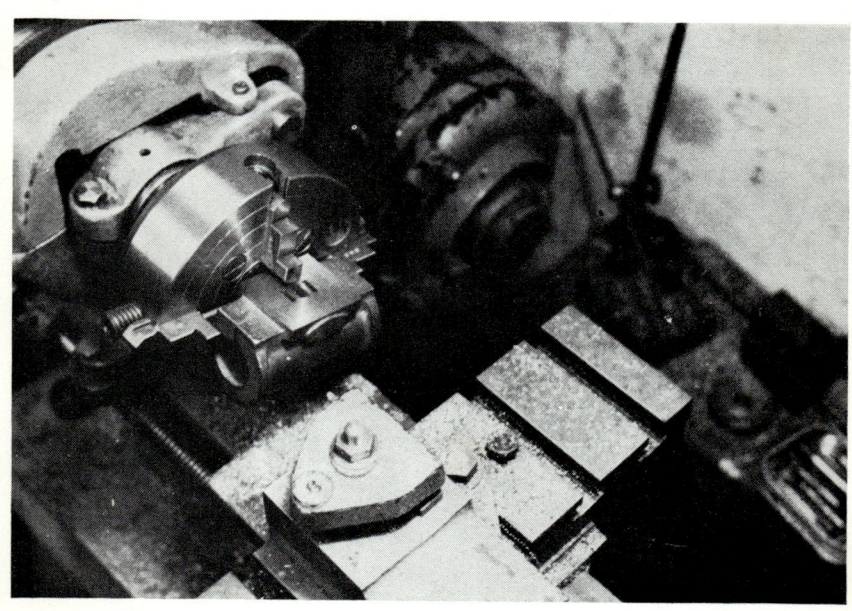

Machining the exhaust flange facing

To bore the cylinder and face the ends, an angle plate bolted to the faceplate was used. The angle plate was made many years ago from a scrap of 3 inch angle and has been worth "a guinea an ounce" ever since. The cylinder was mounted, with a protective piece of paper, on its portface, so that the outside of the flanges ran true. Longitudinal accuracy was taken care of by checking from the faceplate with a try-square.

The cored hole through the cylinder is $\frac{7}{8}$ inch diameter which allows a sufficiency of metal to permit an accurate bore, without the tedium of a multitude of long slow "boring" operations on a light lathe. Mark the end of the cylinder where bore and flange were machined at the same setting; this will be the gland or "front" end when the engine is assembled.

I suggest that we now screw the four-jaw chuck back on the lathe, because many of the remaining components require this facility and it will save some time if we group these operations together.

First we might start on the steam chest. After cleaning with a file, especially the 1 inch by 13/16 inch interior, we can face all the necessary surfaces to size. The important thing here is to notice that difference in thickness ($\frac{1}{4}$ inch and 3/16 inch) on either side of the centre-line. I suggest that you keep the $\frac{1}{4}$ inch dimension fairly full, as any reduction here will limit the clearances available when fitting the slide-valve. The final operation should be to set the gland to run true, face, centre drill and drill 3/32 inch continuing on into the opposite end of the chest. Then open out with a No. 31 drill, through the gland boss only, and ream $\frac{1}{8}$ inch. Finally, open out with a $\frac{1}{4}$ inch drill to a depth of 5/16 inch. Mark out and drill the six 7BA fixing holes and, while you are about it, use these holes to spot through on to the port face of the cylinder, drill tapping size and tap 7BA. Start the taps in the drilling machine so that the threads will be square with the surface.

The steamchest cover looks awkward to hold for machining, but is so accurately cast to size that it can be completed by careful filing. Rub the surfaces on a flat smooth file and finish on emery cloth. If you feel competent, by all means face in the lathe, holding in the four-jaw chuck. When finished to size, clamp to the upper surface of the steam chest and drill through from the six 7BA clearance holes in the latter. Finally, in the middle of the cover, drill a 3/16 inch hole for steam admission.

While the four-jaw chuck is mounted on the lathe, let's have a session of elliptical gland and flange machining. These, luckily, are supplied as neat little gunmetal castings, and I suggest tackling them all together. Clean them up by filing, but do not, for the present, reduce the elliptical part to finished size. The "modus operandi" is, more or less, the same for each of these pieces, so one description will suffice.

Gripping by the elliptical flange, mount the casting in the chuck with the boss outwards and running as true as possible. Hold so that sufficient of the flange is protruding to allow a light facing cut to be taken over it. Face the end of the boss and, if you feel at all uncertain, centre drill the boss so that it can be supported by the tailstock. Turn the boss to diameter and skim the flange at the same time. Now drill the boss and in the case of the glands ream 3/16 inch and $\frac{1}{8}$ inch respectively, while the steam flanges are tapped, 3/16 inch by 40 tpi for the inlet and $\frac{1}{4}$ inch by 40 tpi for the exhaust. Later the glands can be set in their respective positions and the ellipses filed and polished to blend nicely with the gland bosses. The steam flanges should be filed until they give a shape which is pleasing to the eye.

Finally mark out and drill for the fixing bolts and studs.

We might now turn our attention to the piston assembly, and so that we can finish turn the piston on its rod we will make this first. It is a length of 3/16 inch diameter stainless-steel and because it has been precision ground, we should be very careful not to mark or damage its surface. Nothing causes leaky glands quicker than packing which is being abraded by piston or valve rods whose surface is marked by having been gripped, without protection, in chuck or vice.

Make a small split bush from a scrap of brass or aluminium rod, mark it to show the location of No. 1 chuck jaw and split it with a junior hacksaw. A similar bush, made from a piece of square bar, may be necessary if the 5BA thread needs completing in the bench vice.

Hold the 3/16 inch rod by the bush in the self-centring chuck, face the end and turn a $\frac{1}{2}$ inch length to $\frac{1}{8}$ inch diameter; thread 3/16 inch of this to 5BA. Reverse in the chucking bush and face down to a length of $6\frac{5}{8}$ inch from the shoulder.

The piston is a gunmetal casting with cast-on chucking spigot. Clean up with a file. Then mount by the piston in the self-centring chuck with the spigot outwards. Removing the minimum of metal, turn the spigot true and face the side of the piston. Reverse in the chuck, gripping by the spigot, face the piston to $\frac{3}{8}$ inch length, and lightly centre drill. With the tailstock to help keep things steady, turn the outside of the piston to a good 1/32 inch oversize; then with a 1/16 inch wide parting tool, machine the packing groove. Drill the piston $\frac{1}{8}$ inch and with a tiny boring tool, recess the face of the piston to take the 5BA nut.

Reverse the piston in the chuck and face off the remains of the chucking spigot. Clear any burr from the hole in the piston; then, with suitable protection, grip the piston rod by the plain end in the tailstock chuck and gently press the 5BA end through the hole in the piston. Remove the assembly and fit the 5BA nut. Remount by the rod (and bush) in the lathe chuck and face down the 5BA thread to give a neat finish. Then, using the cylinder as a gauge and with a really sharp tool, carefully skim the piston, removing only a gnat's whisker at a time, until it slides without slop into the cylinder. Remove piston and rod from the lathe and put it somewhere where there is absolutely no risk of it being damaged!

Now to the cylinder covers. The top cover is a bit awkward, as it has no chucking spigot. The sequence of operations is as follows. Grip in the chuck by the locating spigot, gland boss outwards. Face the boss to 5/16 inch long. Centre drill and steady with the tailstock. Turn the periphery to $1\frac{5}{8}$ inch diameter and face an annular rim on the cover for the heads of the fixing screws to bed on; then with the tailstock chuck, drill and ream the boss 3/16 inch. Open out to $\frac{3}{8}$ inch diameter for a depth of 5/16 inch, and check that the gunmetal gland fits easily.

Machining front cylinder cover

To turn the 1 inch diameter locating spigot, we will require something to hold the cover so that the rest of our turning will be concentric with the hole in the gland. A chucking mandrel is a simple item to make for this purpose. Do not remove this accessory from the lathe until you have completely finished with it. With the cover mounted on it, face to the thickness shown on the drawing and aim for a really close fit of the cover in the cylinder bore. Make sure you are checking with the right end of the cylinder.

The bottom cover is a simpler version of the previous procedure and requires no detailing.

In each case, while the cover is in the chuck, mark a circle of 1-5/16 inch pitch circle diameter with a vee-tool and on it locate, by centre punch, the position of the 7BA or 5BA clearance holes. Drill these, then use the covers to locate the holes on the cylinder ends. Drill and tap these holes in the cylinder.

Use the cylinder gland to spot for the two 7BA tapped holes in the gland boss. Drill and tap these holes and fit two studs. When fitting studs, remember that the end with the short thread is the 'metal end,' i.e. is screwed into the work. The end with the longer thread takes the nut.

Likewise, the 7BA holes for the exhaust flange can be completed if not already done.

The slide valve and its bits and pieces might well form our next bit of work. As we have the three-jaw chuck set up, let's make the valve spindle first. This is from $\frac{1}{8}$ inch diameter stainless-steel. In order to achieve concentricity and avoid marking the surface of the spindle, it would be worth while making a little bush, as previously described for the piston rod. With things organised, turn down to a bare 3/32 inch diameter for a length of $\frac{5}{8}$ inch, aiming for a fine finish and smoothly dome the end with a fine file. Draw the work out further from the chucked bush and thread the next $\frac{1}{2}$ inch length, 5BA. Stainless-steel can be a little awkward at times, so use plenty of cutting oil. Reverse the spindle in the chuck, face to an overall length of 2-9/16 inch, and thread this end 5BA for a length of 5/16 inch. Finally face down to a final length of 2-7/16 inch, after you have checked the threads.

A session with the four-jaw chuck is now called for; first we will make the valve nut. This is a very simple component, but it is crucial that the 5BA tapped hole be accurately square with the nut surfaces. A tiny gunmetal casting may be supplied for this. In my case I made it from a short piece of $\frac{1}{4}$ inch by $\frac{1}{8}$ inch brass bar. Nevertheless, I set it up in the four-jaw chuck to ensure that the tapped hole was square, and unless you are absolutely sure of your drilling machine, I suggest you do likewise.

Now grip the slide valve in the chuck, face outwards. Lightly skim the surface taking off only sufficent to obtain a good finish. If we take off too much, it may necessitate having to deepen the cavity which is a bit of a nuisance. Machine the sides of the valve by lightly facing, or if only a tiny amount needs to be removed, by filing, until its overall size is, as shown, 21/32 inch long by 9/16 inch wide, with the cavity $\frac{3}{8}$ inch by $\frac{3}{8}$ inch centrally situated.

Clean up the back of the valve with a file and mark the position of the two $\frac{1}{8}$ inch wide slots. Carefully hacksaw the bulk of the metal out. With careful sawing, you will be surprised how little filing will be needed to clear out these slots. In spite of all the fancy machine tools that become available, I still think that the ability to saw so accurately that you can skim a scribed line is one of the most useful skills that a craftsman can develop. If a $\frac{1}{8}$ inch end mill or slot drill is available, you can finish to size, after sawing and filing, with the valve clamped to the vertical slide, or, as in my case, under the tool-post.

With the valve assembly mounted in the steam-chest, there must be clearance under the edge of the valve nut, so that the valve is able to lift slightly off the cylinder portface. In all other respects the nut must be a close sliding fit in the valve slots.

The rod ends are straightforward examples of four-jaw chuck work, but like most apparently simple components, they have got to be made just right. In this case, the threaded hole to take the rod must be square with the cross-hole which takes the wrist pin. One description will suffice for both components. Face each end of the square mild steel bar to finish just over final length. Now set up cross-wise in the chuck and centre drill for the cross-hole; drill and ream as necessary. Mount longitudinally in the chuck and set to run true; centre drill, drill and tap as required; then turn down to finish as shown. Very lightly chamfer the end. Reverse in the chuck and face down to the correct length.

That completes the bits and pieces of the cylinder and we can now assemble with a smear of oil and congratulate outselves on the accuracy of fit which is evident in the way the piston and valve slide to and fro. Do not, for the present, fit any gland or piston packing; we will leave that until the final assembly.

This series of period type engines is, as you will appreciate, developed around the same range of castings and yours truly has machined and built all that has been described. With the kits of castings supplied by Stuart Turners, there is normally included a pair of 1 inch piston rings; whether to fit them or not is a decision you must make. If you intend to work the engine often and if you can bore the cylinder to a sufficient degree of accuracy, then by all means do so. But, in my opinion, there are advantages in using soft graphited packing instead. For example, this packing will hold oil and keep the cylinder from rusting if the engine is only run infrequently; also, any inaccuracies in the diameter of the cylinder bore may be allowed for by turning the piston to fit, if soft packing is used. In building these engines I have used the latter method.

5. Connecting rod, guide, valve operating gear

The connecting rod is a very different type from the 'fish-bellied' examples used on the other engines in this series, but, with care, will not be difficult to make. In fact, the original, known as the 'pitchfork' type, was largely made by the blacksmith with little more than a hammer.

If you feel confident of your sawing and filing ability—and you should by now, you can start with a $6\frac{3}{4}$ inch length of $\frac{3}{4}$ inch by $\frac{1}{2}$ inch bright mild steel. Alternatively, it may be built up with the yoke portion bent from a strip of $\frac{3}{8}$ inch by $\frac{1}{8}$ inch mild steel strip, silver soldered to the upper half of the rod. I imagine that when the sets of castings, etc., become available from Stuart Turner Ltd., it will be the former method that will be suggested, as the built-up technique presupposed the availability of brazing equipment.

To make from the solid, start by marking out the position of the $\frac{1}{2}$ inch diameter hole which forms the root of the yoke. This must be drilled to be exactly in the centre of the width of the rod and I suggest that you make a $\frac{3}{4}$ inch diameter drill bush for this purpose. A 3/16 inch hole in the middle of this bush will be ideal for a pilot hole. Open out carefully to $\frac{1}{2}$ inch diameter then mark out and carefully saw down to this hole to form the two arms of the yoke. Clean up the inside surfaces with files—the more carefully you saw, the less work will be called for with files!

The upper part of the rod is operated upon in a similar manner. Drill two holes to form the rod shoulders; then saw and file to outline.

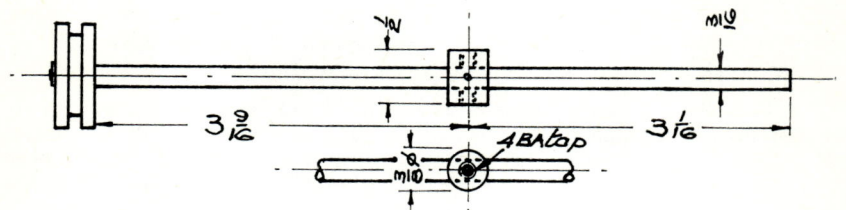

Piston with long rod

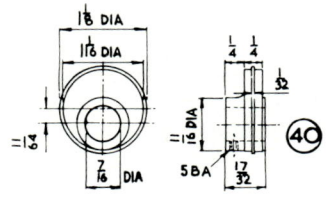

Eccentric sheave

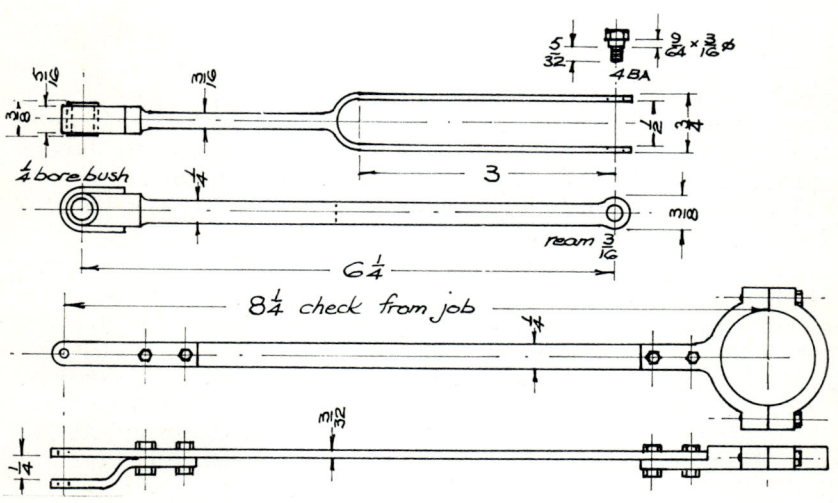

Connecting rod, eccentric rod and sheave

We now have a rod which in one view looks finished but when looked at in side elevation is $\frac{1}{2}$ inch wide all the way down, so some more careful marking out, sawing and filing is called for. The best advice one can give to the beginner when he is faced with these long metal sculpturing jobs, is care and patience. If you really fancy some watchmaking, I have included a drawing of how the crank end of the connecting rod for this, and all the other engines in this series, could be made. It will try your skill and patience to a degree, but think of the satisfaction when you dissemble it in front of your model engineering friends. Will mine be like that? No, I'm afraid not; I have neither the patience nor the skill!

The connecting rod requires a bronze bush which is a simple bit of lathework and two shouldered bolts to screw into the crosshead. The latter are turned from steel hexagonal bar measuring 0·220 inch across the flats, which is a standard size supplied by Stuart Turner. The 4BA thread can, with advantage, be a slightly tight fit in the tapped holes of the crosshead to stop these bolts from slackening back.

The crosshead is an ultra-simple affair on this engine. It consists of a $\frac{1}{2}$ inch length of $\frac{3}{8}$ inch diameter bright mild steel bar drilled and tapped 4BA while set up in the lathe; then—this is the awkward part— it is drilled and reamed 3/16 inch diametrically through to take the piston rod. This 3/16 inch hole must not only closely fit the rod, but must also be exactly square with the axis of the bar. When correctly located on the

Boring the eccentric strap

$3\frac{1}{2}$

$\frac{1}{4}$ ream

7BA clear

$\frac{4}{5}$ ream $\frac{3}{16}$

$\frac{3}{16}$

$\frac{1}{32}$

$\frac{1}{5}$

$\frac{3}{16}$

$\frac{1}{2}$

7BA tap

$\frac{1}{2}$

4BA clear

$1\frac{7}{8}$

$\frac{7}{8}$

matching taper

$\frac{3}{16}$

$\frac{3}{4}$

$\frac{1}{8}$

$\frac{1}{16}$

$\frac{1}{2}$

$\frac{1}{16}$

$\frac{9}{32}$

$\frac{7}{32}$ No.10

$\frac{3}{32}$

$\frac{3}{16}$

$1\frac{1}{32}$ No.10

$\frac{7}{32}$

$\frac{3}{16}$

$\frac{1}{4}$

$\frac{1}{16}$

$\frac{3}{8}$

10BA locking screw

26

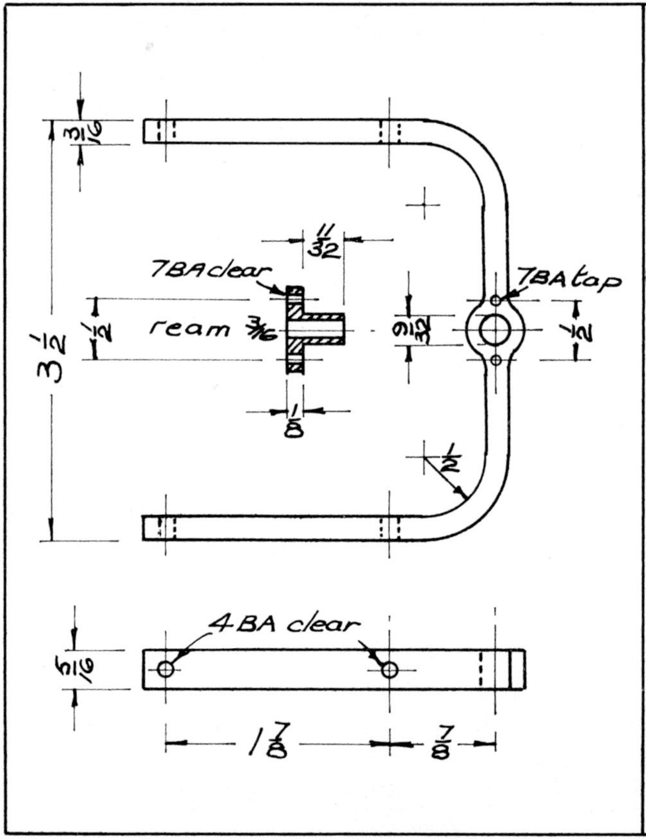

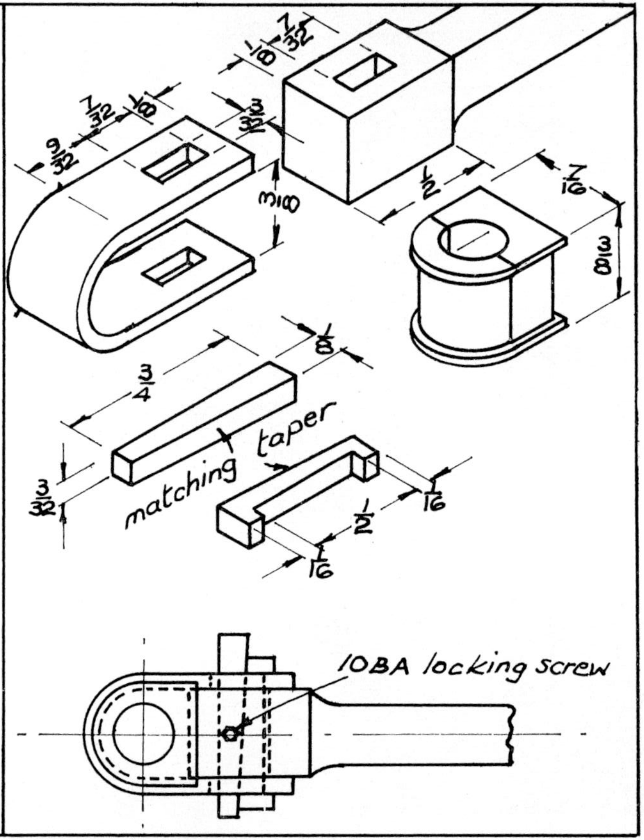

piston rod, drill and fit a 1/16 diameter crosspin to fix both components together.

In place of the usual form of crosshead guide, this early type of steam engine had an extended piston rod, the end of which was located in a bush whose duty was to constrain the piston to move in a straight line. In many such engines, especially if of the horizontal configuration, this bush was replaced by a feed pump so that the piston rod extremity acted as the ram, quite a sensible arrangement if substantially made. The bush is a bit of elementary turning and the material will probably be supplied as a neat little gunmetal casting similar to the piston rod gland. It is machined in exactly the same way.

The support which carries this bush is again most conveniently made from the solid. A casting would save a lot of work, but there is the problem that it might need a bit of bending to get it to fit the individual builder's structure and although the iron that Stuart's use is renowned for its ductility, over-tweaking might not do it any good! The alternative is cutting from the solid and with the connecting rod behind us this job will hold no terrors! Once cut and filed to shape, it is worth while making a bending jig from an odd piece of, say, 1 inch by $\frac{3}{8}$ inch steel bar. Cut it accurately to length, $3\frac{1}{4}$ inch (but check from job); then round two corners to $\frac{1}{2}$ inch radius and finally file a clearance to miss the boss. Locate this boss in the centre of the jig and carefully, with a copper hammer, or at least some protection under your ordinary hammer, tap the two arms round at right angles.

File out the recesses in the columns if not already done, and offer up the support; clean out the recesses so that it is a good fit. Remove the support, mark the positions of the 4BA fixing screws and drill them clearance size; then replace the support checking its position with the piston rod. Spot through into the columns, dis-assemble,—again drill and tap the columns, 4BA.

It is in a position like this that we really begin to realise that building model steam engines is most definitely not like putting plastic kits together!

Now we come to just about the last part of the construction work, the valve operating gear. Start with the eccentric strap. This is a gunmetal casting which after cleaning with a file requires the holes for the 7BA clamping bolts to be drilled tapping size. Now carefully saw the strap in half and face the sawn surfaces by rubbing on a smooth flat file. The 7BA holes may now be tapped or opened out to clearance size as the case may be. Now bolt them together; use a couple of spare 7BA bolts, as during the next few machining operations you may inadvertantly give the heads a clout, and set up the strap in the four-jaw chuck with the bore running as truly as possible. Bore to 1-1/16 inch diameter and recess the bore with a narrow boring tool, rather like a parting tool version of a boring tool. Aim for a really fine finish. Face the outside of the strap, removing about half of the excess thickness of metal but making sure that eventually the clamping bolts will finish in the centre of the strap width. Reverse in the four-jaw chuck; make certain that the faced side will run true by setting it up with some parallel packing between it and the chuck face. Now machine the second side to the finished width of $\frac{1}{4}$ inch. Clean off burrs and drill a tiny oil hole. Finish the protruding arm on which the rod is to be bolted by filing or milling.

The eccentric sheave is turned from a short length of bright mild steel. Face; then use the strap as a gauge for turning the 1-1/16 inch diameter and the raised rib. When this part has been completed, reverse the sheave in the chuck and face the opposite side. Remove from the lathe and with steel rule and scriber, draw a line diametrically across the faced surface. From the centre of the sheave, which will be evident from the turning marks, measure out 11/64 inch and lightly centre punch. Reset in the four-jaw check and set the centre punch mark to run true. Centre drill, open out and ream 7/16 inch; then turn the outside down to 11/16 inch diameter. Face to an overall length of 17/32 inch with the sheave $\frac{1}{4}$ inch long; likewise the boss and a tiny clearance boss at the opposite end. Drill and tap for the 5BA grub screw to lock the sheave to the crankshaft.

The eccentric rod is a straight length of $\frac{1}{4}$ inch by 3/32 inch bright steel bar. The centre distance should be $8\frac{1}{4}$ inches, but play safe by checking this when the engine is assembled. A bridle piece is made from the same steel strip and bolted to the end of the rod to make a yoke within which the end of the valve spindle fits.

Of all the engines in this series so far, the "REAL" has the simplest arrangement of valve operation, the eccentric rod being directly connected to the valve rod. This is an obvious advantage in building and, in my opinion, makes this the simplest engine for the real beginner to make a start on.

6. Assembly and running

I expect you have assembled the engine as you worked through the making of the various parts. If you have been careful, you will now have everything working smoothly but without any sloppiness. As no cylinder or gland packing has been fitted, we know that any tight spots that do exist are not due to this, but to metal to metal contact which we must clear before proceeding further.

Two things which require a decision are whether to fit cylinder drain cocks and lubricators. The former, part No. 196/0, requires a hole to be drilled No. 30 or 3·30 mm and tapped 5/32 inch by 40 threads per inch at each end of the cylinder block. At the end of its stroke, the piston will cover these drain holes, so, with the corner of a small square or triangular file, make a groove from the tapped hole to the end of the cylinder block.

An alternative to the straight nose cock is to fit union cocks (No. 200/1) and run some $\frac{1}{8}$ inch outside the diameter copper pipe down to a drip tray below the base of the engine. This method is to be preferred if you intend running the engine on steam as there is then no risk of getting a face full of steam or boiling water.

Fit the drain cocks with a tiny smear of pipe jointing compound and use an aluminium washer to get the cock to stand the right way up when screwed home. These washers are available from Stuarts and are a great help for the final assembly of engine and boiler fittings.

Oil cups (No. 183/1) do a lot to give a n engine a finished appearance and look nice fitted to the main bearings, connecting rod and eccentric strap. The size suggested for this engine requires a $\frac{1}{8}$ inch by 40 tpi tapped hole, tapping size drill No. 40 or 2·50 mm.

The length of the cylinder block is 2-9/16 inches and if we deduct from this the stroke, 2 inches; the piston length, $\frac{3}{8}$ inch; and the cylinder cover spigots, $\frac{1}{8}$ inch; we are left with a total clearance of 1/16 inch which means a clearance between piston and cover at each end of 1/32 inch. By careful measuring and checking, ensure that you have this clearance at each end of the stroke before you finally peg the crosshead to the piston rod.

With this work complete I suggest that you now dismantle the engine for final cleaning and painting. You have complete freedom of choice here, but don't let the colours be too garish; after all this was a working engine, not part of a fairground! The connecting rod and the columns may be left bright, although the latter may be blued or painted black. Actually the columns on my version and in the drawings are as simple as possible, but there is no reason why you should not express your artistic self in turning these; the original builders of these engines had no inhibitions when it came to decorating them.

While the paint is drying, make the cylinder and steam chest gaskets. Two of each are required, the steam chest ones being both the same while the cylinder gaskets differ in the number and size of the holes. They are made from strong brown paper; first marking them out as a drawing board exercise, and then cut them out with scissors. Now well smear them with cylinder oil, rub it in with your finger and fit them in place on the cylinder covers and on the steam chest and its cover. With a scriber make a hole from the paper side through the metal cover. Push the tapered scriber right in with a twisting motion so that the paper is sheared through as the taper on the scriber meets the sharp edge of the drilled hole. Leave them in place and assemble the cylinder parts. Leave the steam chest cover off until we have set the valve.

With everything assembled, turn the engine over in the direction in which it is desired to run. With this particular design it can be either way, the limiting factor not being the crosshead guide design—as is usually the case—but the direction of the driving belt from the engine pulley, it being normal practice to have the lower part of the belt pulling.

For this description let us imagine that the engine is required to rotate clockwise *when viewed from the back*. Working from the back will make it easier to see the valve.

Turn the engine over via the flywheel and watch the travel of the valve. Alter the valve position until it uncovers the steam port by the same amount at each end. Do this by taking the bolt out that joins the end of the eccentric rod to the valve spindle and rotating the valve spindle the required direction. Get the port opening as near the same at each end as possible. Now slacken the grub screw locking the eccentric sheave to the crankshaft and turn the crankshaft round, in the direction of rotation, until the piston is at the bottom of its stroke; then, with the crankshaft stationery, rotate the eccentric sheave *in the same direction* until the valve has gone to the bottom of its stroke and has started to rise up again. Now watch carefully and as soon as the valve starts to uncover the lower steam port, *stop*, and lock the eccentric sheave to the crankshaft again. Continue turning over the crankshaft via the flywheel and watch the repeat performance taking place at the other end of the cylinder. By the time the piston has reached top dead centre, the valve will have gone right to the highest point of its stroke, will be returning and will just be uncovering the top steam port. If both ends of the chest have the same conditions, congratulate yourself and fix the steamchest cover, not forgetting the oiled paper gasket; if there is a discrepancy in the valve positions split the difference by slightly altering the position of the sheave on the crankshaft. If the error is great, do a careful check of port and valve sizes and the throw of the eccentric sheave.

If your engine is for mantlepiece decoration, mount it on a nicely polished piece of hardwood: rich mahogany from an old piece of furniture looks very pleasing. If it is to be steamed, a displacement lubricator (No. 185/1) should be fitted to a hole drilled and tapped in the steamchest wall; alternatively the combined lubricator and stop valve (No. 155) may be used. Mount the engine, in this case, on a hard water resistant surface set in a metal tray to collected odd drips of oil and water.

I hope you have enjoyed building this engine and am sure that it will give you many years of fun.

Cylinder being bored, mounted on angleplate bolted to faceplate. Lathe changewheels used as balance weights

Machining surfaces of steam-chest

2⅝″

The perfect lubricator for steam engine cylinders

One of the many accessories listed in our Model Catalogue, price 35p (post free in the UK).

STUART TURNER LTD.

HENLEY-ON-THAMES
OXON., RG9 2AD, ENGLAND

SIMEC

The Stuart International Model Engineers' Club was formed at the end of 1971 to satisfy the needs of those model engineers who either want to " go it alone " or who have no club facilities locally. SIMEC now has several hundred members throughout the world.

The Members' Directory also lists particular engineering skills which is obviously of great value to beginners or those seeking help or advice. For details of membership write to:—The Secretary (SIMEC), c/o. Stuart Turner Ltd., Market Place, Henley-on-Thames, Oxon., RG9 2AD, England.